A Patient's Guide to

CANCER

Understanding the Causes
and Treatments of a
Complex Disease

John F. McDonald

For information about this title or
to order other books and/or electronic media,
contact the publisher:

ЯP
Raven Press, LLC
Atlanta, GA

Raven Press
1445 Woodmont LN NW #1373
Atlanta, GA 30318

ISBN:
979-8-9876276-0-0 (paperback)
979-8-9876276-1-7 (ebook)

Front book cover images:
LumineImages/Shutterstock.com; Flywish/Shutterstock.com;
Christoph Burgstedt/Shutterstock.com

Back book cover images:
Christoph Burgstedt/Shutterstock.com

Printed in the United States of America

*This Book is dedicated to
my fellow cancer survivors,
past, present and future*

Table of Contents

Preface

What is cancer?

This is a frequently asked question, not only by the general public but perhaps most urgently, by newly diagnosed cancer patients, their families and friends. The question is also asked by the thousands of clinicians and research scientists who seek to better diagnose, treat and eventually eradicate the disease. As it turns out, there is no single answer to this question because cancer is a complex, multifaceted disease that manifests itself on many different levels.

While nearly everyone recognizes cancer as a devastating disease that affects large numbers of individuals on a yearly basis, from the patient's perspective it's much more. Newly diagnosed cancer patients find themselves suddenly and unexpectedly facing a reality laden with intense emotions of both fear and hope, not to mention the physical discomfort associated with the disease and/or by prescribed treatments. From the clinician's perspective, cancer brings with it the immense challenge of choosing the most appropriate treatment for individual patients from an ever-expanding list of options.

From the perspective of research scientists, the disease also manifests itself on multiple levels. On the cellular level, cancer presents itself as the uncontrolled growth and replication of abnormal cells. For cancers derived from solid tissues, this uncontrolled growth typically culminates in a localized solid mass of abnormal cells, called a **primary tumor**. Cancer cells may slough off from this primary tumor and spread (**metastasize**), typically via the circulatory or lymph system, to other body tissues, establishing secondary or metastatic cancers. On the sub-cellular level, the disease is associated with the abnormal structure and/or regulation of molecules (proteins, nucleic acids, metabolites) that control essentially all aspects of cell function. Because cancer cells replicate and pass their undesirable characteristics on to **progeny cells** (cells make more cells when necessary and these cells are called "progeny" or "offspring"), the origin and progression of the disease must be linked to information (or perhaps more accurately, "misinformation") stored in the cancer cells' hereditary material (DNA) that is passed on to progeny cells.

The primary goal of this book is to empower you, if you are a cancer patient, and your family, by demystifying the processes underlying the disease and explaining how recent discoveries about these processes are translating into the development of promising new approaches to cancer diagnosis and treatment. While this book is intended to be generally accessible to non-scientists, additional references/links are provided at the end of each chapter for those interested in acquiring more in-depth coverage of individual topics. In addition, key terms are highlighted throughout and are formally defined in the Glossary at the end of the book.

Certainly, no one wants to develop cancer, but if you do, it's important to know that it's no longer a mandatory death sentence. As a cancer patient, the more you understand about the disease and recent scientific breakthroughs, the more prepared you will be to discuss optimal and appropriate therapies with your physician(s), and ultimately, make informed decisions about your future life.

Animal Cell

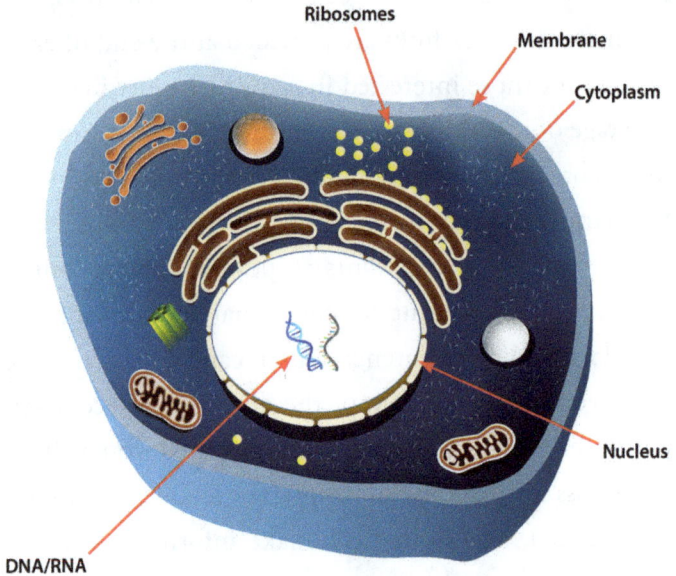

Figure 1- The basic structure of animal cells. The cell is the basic building block of living organisms. It has three main parts: the **cell membrane**, the **nucleus**, and the **cytoplasm**. The cell membrane surrounds the cell and controls the substances that go into and out of the cell. The nucleus is a structure inside the cell that contains the cell's DNA and is where RNA is made. The cytoplasm is the fluid inside the cell containing other tiny cell parts that have specific functions. The **ribosomes** are where proteins are made. The cytoplasm is where most chemical reactions take place. The human body is comprised of more than 30 trillion cells Designua/Shutterstock.com).

Chapter 1
An Introduction to Molecular Biology

DNA – the blueprint of our body

Think of **DNA** (<u>d</u>eoxyribo<u>n</u>ucleic <u>a</u>cid) as a blueprint or a database containing all of the information necessary to make you what you are both structurally (*i.e.*, the cells comprising all of your various organs and tissues) and functionally (*i.e.*, all metabolic, enzymatic, replicative functions, *etc.*). Every cell (Figure 1) in your body contains the entire complement of this information, safely stored within a subcellular structure, called the **nucleus**[1]. However, only a subset of the information stored in your DNA is required for the specific structures and functions associated with your various types of cells. (Think of it like this: as a skyscraper is being built, the workers constructing the foundation aren't focused on the parts of the blueprint that address the

---

1   The sole exception is mature red blood cells that lose their nuclei and DNA during their development and, as a consequence, do not divide/replicate, and thus, have a limited lifespan.

construction of the roof. Only a subset of the information contained within the blueprint for that skyscraper will be utilized by those workers.)

For example, even though all of the information necessary for both liver and brain cell functions are stored in the DNA of both cell types, the detailed information for liver cell function is not utilized in brain cells and *vice versa*. Rather, the subset of the information stored in DNA necessary for cell specific functions is transferred to an intermediary molecule, called **RNA** (ribonucleic acid), that then transmits this cell-specific information to the cell's protein factories (called **ribosomes**) where the information is used to construct all of the proteins necessary for specific cell functions.

Thus, if you were to collect and compare DNA from your liver cells and your brain cells, it would be identical, but the RNA collected from your liver and brain cells would be quite different, reflecting the specific information needed to execute the specialized functions associated with each cell type. Disruptions in this orderly flow of information from DNA to RNA to protein (Figure 2) is the basis of altered cell function(s) and, if these altered functions are detrimental to the normal operation of cells, they may manifest as any number of cellular abnormalities – including cancer!

Information Flow in Cells

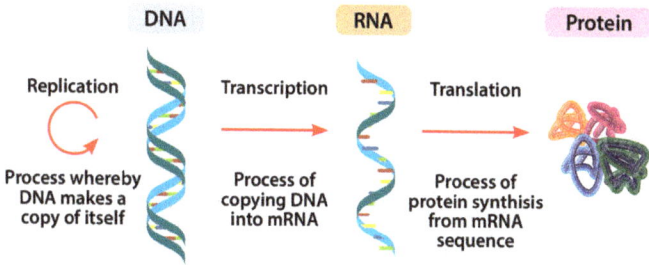

Figure 2- Information flow in a cell. The subset of the information stored in DNA is copied for transmission to progeny cells by a process called replication. In individual cells, the information stored in DNA that is necessary for cell-specific functions is transferred to an intermediary molecule called mRNA by a process called transcription. This cell-specific information is then transferred to the cell's protein factories called ribosomes by a process called translation. The information is used to construct all of the proteins necessary for specific cell functions (BigBearCamera/Shutterstock.com).

Disruptions in the flow of information in the cell may occur on multiple levels

Let's think for a minute about where these "disruptions" in information flow could arise. One possibility is that there may be a change in the information stored in the DNA database itself. These changes in DNA, called **mutations**, would then be transferred to RNA and ultimately result in the synthesis of abnormal proteins potentially leading to atypical cellular function(s).

For example, one of the characteristics that distinguish cancer cells from normal cells is uncontrolled DNA **replication** and cell division leading to the formation of tumors. Normally, the replication of cells is a highly regulated process that is turned on when appropriate, *e.g.*, during embryonic development when the fertilized egg undergoes rapid replication to generate the trillions of cells needed to comprise a complete organism. Once the appropriate number of cells are generated during embryonic development, the process of cell replication is largely repressed and cells stop dividing and proceed to differentiate into the specialized structures and functions associated with our various organs.

One of the proteins involved in repressing cell replication of fully developed retinal cells in the eye is a protein, called RB-1. It has been shown that mutations in the DNA encoding region of this protein (*i.e.*, the RB-1 **gene**[2]) can result in loss of its repressive function. This loss results in rapid cell division within the context of otherwise highly differentiated retinal cells, leading to formation of a cellular mass or tumor in the retina, called retinoblastoma (Figure 3).

---

2 In this context, the term **gene** is typically defined as that segment of DNA that encodes the information needed to produce a single protein. We each carry two copies of every gene- one inherited from our mother, one from our father. Sometimes the loss of function associated with a mutated gene/protein can be compensated for by the other, non-mutated copy of the gene. In cases like this, we refer to the mutated copy of the gene as "recessive" because the detrimental effect associated with the mutation is being masked by the normal copy of the gene. In order for recessive mutations to have their negative effect, both copies of an individual's gene would have to be mutated.

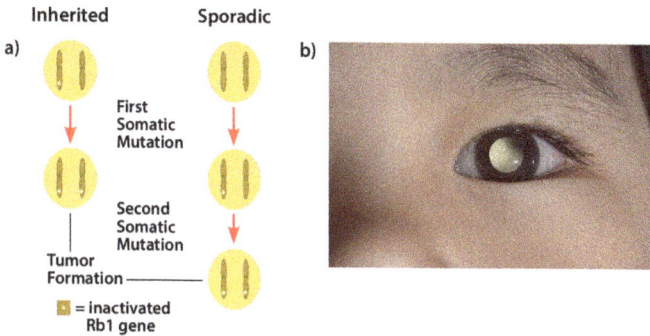

Figure 3- Retinoblastoma. The RB-1 protein is involved in repressing cell replication of fully developed retinal cells in the eye. (a) Mutations in each of the two RB-1 genes encoding the RB-1 protein can result in loss of the protein's repressive function resulting in rapid cell division and leading to (b) formation of a cellular mass or tumor in the retina called retinoblastoma. If a child inherits a mutant RB-1 gene, then only a single mutation in the normal gene can lead to the cancer. In the sporadic form of the disease, there is no inherited pre-disposition for cancer. Rather, two independent mutations in the normal RB-1 gene would be required for tumor formation (ARZTSAMUI/Shutterstock.com).

Our cells are equipped with DNA repair mechanisms to prevent excessive accumulation of mutations

If you think about it, the process of mutation might seem to be a bad thing and, if allowed to get out of control, could result in so many disrupted non-functional cells that life would be impossible. This is certainly true. As a preventative measure, our cells are equipped with an elaborate system of **DNA repair mechanisms** (encoded by **DNA repair genes**). The goal of our DNA repair mechanisms is to fix any mutated or otherwise damaged DNA, resulting in reducing the number of disrupted cells and thereby maintaining normal cell function. Although this repair process is highly effective, it is not perfect –a few mutations are inherited by progeny cells every round of replication.

In some cases, the level of mutation/DNA damage is so severe that it is beyond repair. In cases like this, our bodies have developed a back-up mechanism whereby cells associated with severely damaged DNA are instructed to self-destruct. The act of self-destruction prevents replication of those cells carrying severely damaged or highly mutated DNA. This back-up surveillance system is called, **apoptosis**, or "**programmed cell death**." This will be discussed further in Chapter 3.

Cancer may be caused by a disrupted flow of information in addition to mutations

In addition to mutation, another way by which the normal flow of information in cells can be disrupted may not involve a change in the information encoded in DNA at all, but rather a disruption in the flow of information from the DNA to RNA and to protein. For example, we previously considered the negative effect of a **loss-of-function mutation** in the segment of DNA or gene encoding the RB-1 protein and how this can lead to abnormal cell growth and tumor formation. Suppose that rather than a mutation in the RB-1 gene, the transfer of the correct information from the DNA to RNA to protein was suddenly blocked or significantly reduced.

The net effect would be the same, *i.e.*, cell replication would no longer be repressed, leading to increased cell division and subsequent tumor formation. In the former case, the flow of information from DNA to RNA to protein is normal, but the information being transmitted from DNA is mutated and incorrect. In the latter case, the information encoded in the DNA is correct but it is not being transmitted to the RNA and protein as it normally should. This would be a disruption in the regulated flow of information from the DNA database, rather than a mutation in the DNA database *per se*. Both types of events may occur independently or together, leading to the disrupted flow of information associated with cancer.

Although the term, **gene,** is typically defined as that segment of DNA encoding a type of RNA (called **messenger RNAs or mRNA**) that carries the information needed to synthesize proteins, some genes produce RNAs that never get translated into proteins. Current evidence suggests that the majority, if not all, of these **"non-encoding RNAs"** are playing a regulatory role in the cell, contributing to the properly controlled transfer of information out from the DNA database. It is estimated that the human genome is comprised of around 25,000 protein-encoding genes, but the segments of DNA encoding regulatory RNAs may be much larger.

The bottom line is that all of the DNA we carry in our nuclei (collectively referred to as our **genome**) encodes all of the information needed to make us what we are.

Additional References and Links

Your Genome (yg) 2022. A website produced by the Public Engagement team and scientists at the Wellcome Genome Campus near Cambridge in the United Kingdom. (https://www.yourgenome.org/).

An Interactive Introduction to Organismal and Molecular Biology by Andrea Bierema. https://openbooks.lib.msu.edu/isb202/chapter/introduction-to-molecular-biology/

Chapter 2

The Molecular Basis of Cancer

Cancer was initially believed to be caused by mutations in relatively few genes

Now that we know how misinformation can be transferred (or not transferred) from DNA (genes) to RNA and proteins, the next question is which and how many genes are involved in the transfer (or misregulated transfer) of the aberrant information necessary to develop cancer?

Initially, it was believed that cancer may be the result of disruption in the information flow of relatively few genes. This initial belief derived predominantly from studies (starting in the early 1900s) carried out on cancer-causing viruses. Surprisingly, many of these viruses were eventually found to carry mutated and/or dysregulated versions of a few normal human genes that were subsequently shown to be responsible for the ability of these viruses to cause cancer.

The unexpected finding that aberrant versions of human genes carried by some viruses could cause cancer suggested to early cancer researchers that the corresponding normal versions of these genes, carried by all of us, might also have the potential to cause cancer if disrupted by mutations or mis-regulation.

Cancer biologists have learned a lot more about the molecular basis of cancer since the 1960's. It is currently estimated that there are between 500-1000 genes that, when disrupted, may contribute to the onset of one or more types of cancer. These genes are today collectively referred to as **cancer driver genes** because they may significantly contribute to, or "drive," the onset and/or progression of cancer.

There are multiple molecular pathways leading to cancer

Interestingly, it has been observed that disruption in the information flow of relatively few cancer driver genes is universally associated with the onset/progression of all, or even multiple, types of cancer. In other words, the majority of cancer driver genes that have been identified are most often associated with the onset/progression of *specific* cancer types. Why is this? Growing evidence suggests that the primary reason may lie in the fact that our genome is a highly integrated network of interacting genes and, as a consequence, there appear to be alternative molecular paths (associated with different cancer driver genes) by which normal cells can transform into different (and even the same!) types of cancer.

As an analogy, think of the highly integrated system of roads and highways in the United States. I live in Atlanta. If I want to drive to New York City, I have many options (probably thousands). Normally, I would take the most direct and fastest route, which would be I-85 running northeast to connect with I-95 heading north all the way to NYC. Obviously, depending on where you live, the route you would take to get to NYC would differ but may (*e.g.*, if you live along the east coast) or may not (*e.g.*, if you live in Chicago) involve some of the same roadways.

Think of NYC as cancer and the various possible starting cities as the variety of normal cell/tissue types of which we are composed. Can you get a feel for why the disruptions of pathways and driver genes leading to the emergence of cancer in different tissues can be quite different? Similarly, just as there are alternative routes to NYC from Atlanta, there may be alternative molecular pathways (and associated driver genes) leading to the *same* cancer type. This underlying molecular complexity helps explain why it is so difficult to not only treat cancer on the molecular level, but to accurately diagnose the disease using **molecular biomarkers** (molecules that can be used as an indicator of a particular disease state, like PSA for detecting prostate cancer).

Sometimes even when a molecular pathway leading to cancer is correctly identified and inhibited by an appropriate therapeutic drug, the tumor growth may be initially arrested, but cancer cells may find an alternative molecular pathway in order to resume tumorous growth (*e.g.*, similar to a detour that would allow me to continue my trip to NYC if a section of I-95 had been destroyed). This molecular scenario is one way in which cancers successfully treated initially may, in time, recur.

Additional References and Links

What is Cancer? Bozeman Science video: https://www.youtube.com/watch?v=UopUxkeC4Ls

Animated Introduction to Cancer Biology by Cancer Quest-Emory University. https://www.youtube.com/watch?v=46Xh7OFkkCE

The Molecular Basis of Cancer 4th edition by John Mendelson, Peter Howley, Mark Isreal, Joe Gray and Craig Thompson. Elsevier Press, 2015.

Chapter 3
Current Approaches to the Treatment of Cancer

Surgery is the oldest and still the most common treatment for cancer

The idea of surgically removing cancerous tissue from patients can be traced as far back as ancient Egypt (reference to removal of breast tumors is found in the "Edwin Smith papyrus,"1600 BC). The challenge facing surgeons then, as now, is how to remove as much of the tumor as possible without unnecessarily damaging the patient in the process. Successful cancer surgery involves high technical precision in the surgical procedure and in the determination of the physical borders, or margins of the tumor (**tumor margins**). This multi-pronged approach minimizes the probability that cancerous tissue will remain after the surgical procedure is complete. Modern cancer surgery incorporates a number of technological advances that facilitates attainment of both of these goals and, as a consequence, significantly improves patient outcomes.

For example, today surgeons are often able to utilize a procedure, called **laparoscopy,** to access tumors localized in the abdomen or pelvis without making large incisions. In this procedure (sometimes called, "keyhole," or **minimally invasive surgery**), the surgeon makes one or more small incisions, allowing insertion of a **laparoscope** (a fiber optic instrument) and small surgical tools to visualize and remove the cancerous tissue (Figure 4). Laparoscopic surgery is often a preferred approach when appropriate because it is generally associated with reduced postoperative pain and complications when compared with comparable **open surgical techniques.**

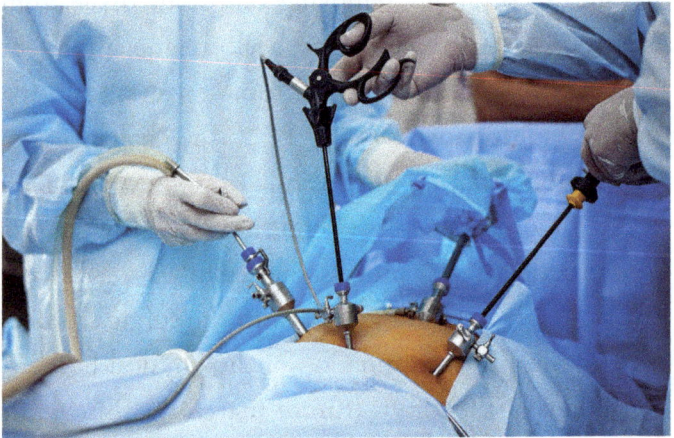

Figure 4- Laparoscopic Surgery. Shown are common instruments inserted into the body during laproscopic surgery. Carbon dioxide gas is used to inflate the abdominal region so there is a better view and more room to work. The laproscope has a light source and a video camera so the doctor can view the internal organs. Other tools generally include a "grasper" for grabbing hold of tissues, a needle holder for suturing and various tools for cutting (Flywish/Shutterstock.com).

A recent extension of minimally invasive surgery is **robot-assisted laparoscopic surgery (RALS)**. RALS is not available at all surgical locations due to the relative high cost of installing the systems and the current limited availability of appropriately trained surgeons. However, robotic assisted surgery is, nevertheless, rapidly expanding because robotic systems often incorporate sophisticated imaging and precise articulated instruments that significantly improve visibility, as well as, the precision of surgical manipulations.

Advanced imaging technologies enhance the effectiveness of modern cancer surgeries

Advanced imaging technologies have also played a significant role in improving the effectiveness of standard cancer surgeries by providing the precise location and extent of tumors prior to and, in some cases, during, surgical procedures. For example, **computerized tomography (CT)** scanning is a technology that generates a series of X-ray images taken from different angles around the body. When these images are combined with specialized computer programs, cross-sectional images (slices) of body parts are generated that reveal the location and extent of cancerous tissues if present (Figure 5). Oftentimes, CT scans are combined with oral ingestion or intravenous injection of dyes to improve contrast levels and the resolution of the CT images.

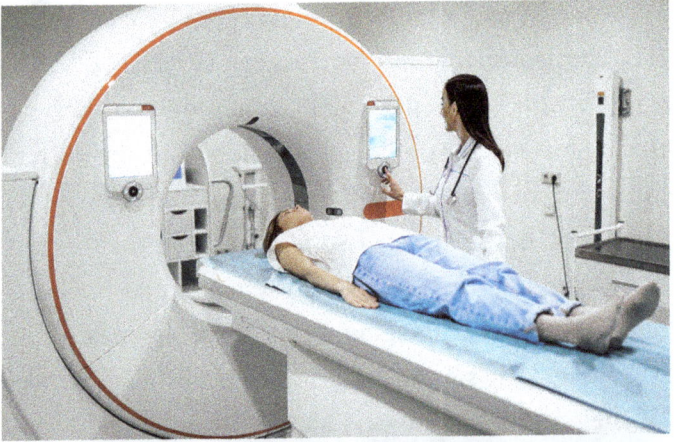

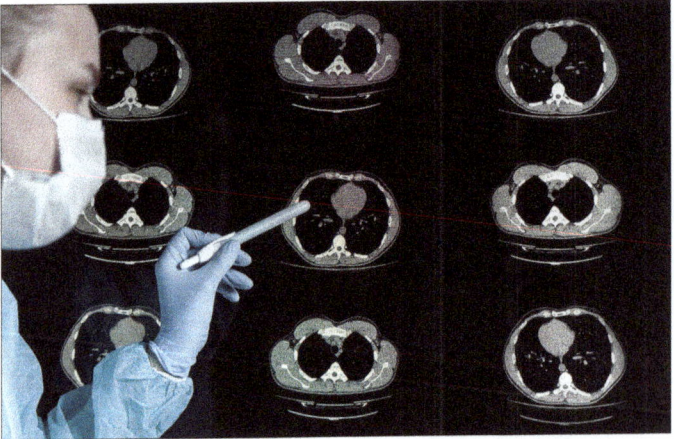

Figure 5- CT scan (computed tomography scan, aka CAT scan). In this procedure, the patient lies on a table that moves through the CT scanner. As the table moves the scanner rotates around the patient. During each rotation images of thin slices of the body are taken. These images are combined with special computer programs in order to visualize the location and extent of cancerous tissues. CT scans are painless and most often an outpatient procedure taking 10-30 minutes (Shyntartanya/shutterstock. com; My Ocean Production/Shutterstock.com).

A second non-invasive imaging technology called **magnetic resonance imaging** (**MRI**) creates generally more detailed images of internal body structures, including cancers if present, by using a large tube-shaped machine that creates a strong magnetic field around the patient (Figure 6). These magnetic waves are combined with pulses of radio waves that, when analyzed by specialized computer programs, generate highly accurate images of body parts. MRI scans are more expensive and take longer to perform than CT scans, but they are the preferred method of detection and localization for some cancers. Unlike CT scans, they are not associated with exposure to radiation.

A third imaging technology called **PET** (**positron emission tomography**) scans focus on detecting metabolic changes characteristic of growing tumors. This methodology requires the introduction of low levels of radioactive material. Metabolic changes are often detectable at very early stages of tumor development, and PET scans may provide physicians with valuable information not available through CTs or MRIs alone. Although quite sensitive, PET scans can sometimes generate **false positive** (indicating a positive signal when cancer is not present) or **false negative** (not showing a positive signal when cancer is present) results. Oftentimes, these alternative imaging technologies are used in combinations to provide the physician with optimal information.

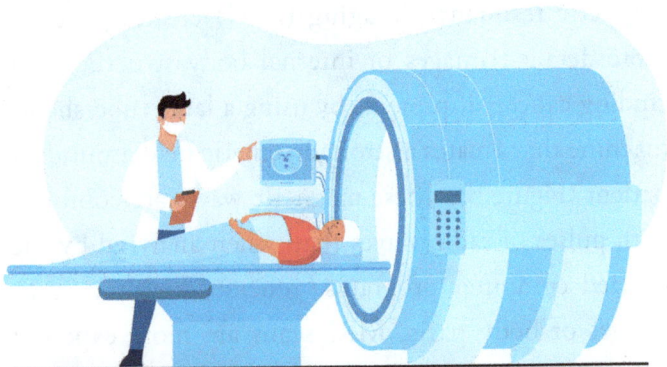

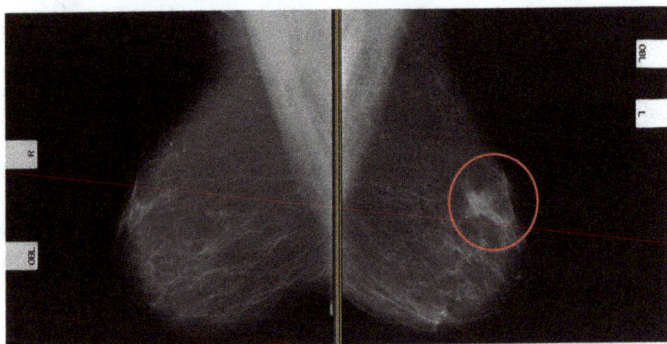

Figure 6- Magnetic Resonance Imaging (MRI). A medical imaging technique that uses a large tube-shaped machine to produce a magnetic field and computer-generated radio waves around the patient to create detailed images of the organs and tissues. Shown is an MRI of breast tissue revealing a tumor in one of the breasts (Hangouts Vector Pro/Shutterstock.com; luckykdesignart/Shutterstock.com).

In addition to its role in imaging, radiation can be used to treat cancers as well

Interest in the use of radiation for the treatment of cancer emerged almost immediately after X-rays were first discovered by the German physicist, W.C. Roentgen, in 1895. For example, the French physician, Victor Despeignes, published a paper in 1896 reporting that a week-long exposure of a stomach cancer patient to X-rays resulted in a significant reduction in both pain and tumor size. Although Despeignes' patient eventually died of the disease, interest in the use of radiation as a treatment for cancer continued into the early 20th century and sometimes was associated with dramatic success. In 1910, the Swedish physician-physicist, Thor Stenbeck, reported that he was able to cure a type of skin cancer (basal cell carcinoma), with small daily doses of radiation.

Today, the killing of cancer cells by radiation utilizes what is called, **ionizing radiation**. This type of radiation forms ions (electrically charged particles) that can severely damage the DNA of the cells comprising tissue that the radiation passes through. As briefly discussed in Chapter 1, severe DNA damage typically induces the somewhat delayed process of programmed cell death (apoptosis) that is initiated when the incurred DNA damage cannot be resolved by the cell's DNA damage repair mechanisms. It is for this reason that the induced death of cancer cells by radiation is often not immediate; it is detectable only several (sometimes many) days after the therapy is initially administered.

As you can well imagine, the downside of radiation is that it may damage normal cells as well as cancer cells. Thus, the goal of modern radiation therapy is to maximize the radiation dose to cancer cells while minimizing exposure to normal cells.

There are two ways to target radiation to the location of the cancer, one external (radiation source outside the body) and one internal (radiation source introduced into the body) (Figure 7).

External beam radiation is the most common type of radiation therapy in current clinical practice. It involves the aiming of high energy particles, generated from a radiation source outside the body, to the location of the tumor (assisted by the imaging technologies discussed above). Subtypes of external beam radiation are defined by the particular type of radiation particle used in the treatment – photons, electrons or protons. **Photon and electron radiation therapies** are commonly available at most radiation centers while **proton therapy** requires more highly specialized and costly equipment. It is not as commonly available at the present time.

What's the difference between these types of radiation particles? Well, if you've had an X-ray of your chest, legs, arms *etc.*, you've already been exposed to **photon beam radiation** although, as alluded to above, much higher energy photon beams are used in cancer therapy.

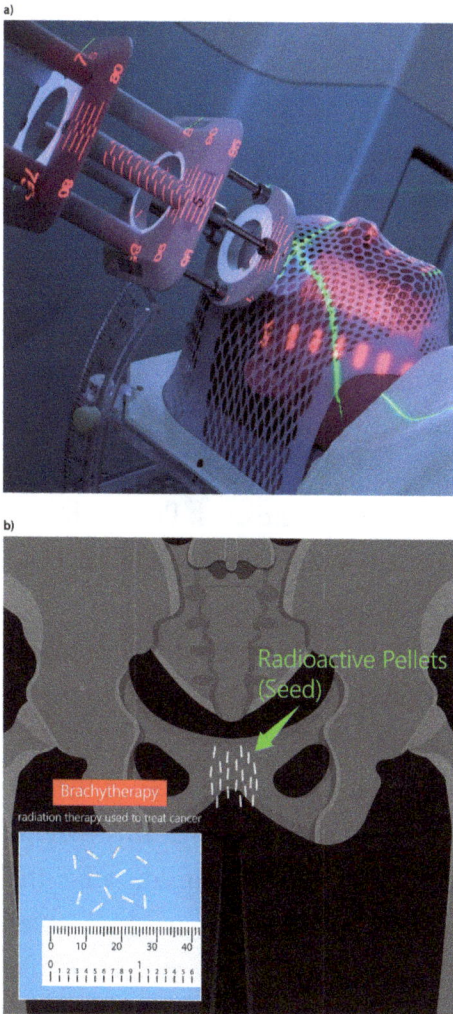

Figure 7 - External and Internal Radiation Treatment. There are two ways to target radiation to the location of the cancer: external (radiation source outside the body) and internal (aka, "Brachyotherapy"- radiation source introduced into the body). Shown are examples of (a) external radiation in the treatment of brain cancer and, (b) internal radiation (implantation of radioactive seeds or pellets) in the treatment of prostate cancer (Mark_Kostich/Shutterstock.com; Pepermpron/Shutterstock.com).

Various imaging technologies (see above) have been developed in recent years to help ensure that the photon beam is targeted precisely to the tumor, thereby reducing damage to adjacent normal healthy tissue. **Electron beam therapy** is not able to penetrate tissues as deeply as photon therapy. Therefore, it is typically used only to treat cancers on or close to the surface of the skin. **Proton beam therapy** is another form of radiation therapy that can be delivered with exceptionally high precision. When proton beams are precisely targeted to tumors high energy is released to the cancer cells with minimal damage to surrounding tissues. Proton therapy is a preferred approach when tumors are located adjacent to critical normal tissues (*e.g.*, in the treatment of tumors of the brain or spinal cord or in the treatment of tumors in children where the goal is to minimize damage to growing healthy tissues).

Less common than external beam therapy is **internal radiation** or **brachytherapy**. This therapy involves the delivery of radiation from inside rather than outside the body. These radioactive sources are sealed in catheters, or small "seeds," that are directly implanted into the tumor site (Figure 7). In brachytherapy, the radiation is being physically focused on the area of implantation, so higher levels of radiation can be achieved. Seeds with lower doses may also be used when the intent is to leave them implanted for extended periods of time. Another positive aspect of brachytherapy is that any negative side effects associated with the radiation treatment are usually limited to the area being treated.

The particular type of radiation therapy selected is dependent on the type of tumor being treated. For example, for some early-stage cancers (*e.g.*, some types of skin cancer, cancers of the prostate, cervix and lung), radiation therapy alone may be appropriate and effective. In the majority of cases, however, radiation therapy is considered most effective when employed in combination with surgery and/or chemotherapy.

Chemotherapy is often an essential complement to surgery and/or radiation in the treatment of cancer

Although, it may seem intuitively obvious to you that the physical removal of cancerous tissue by surgery or radiation induced cell death are highly effective approaches for treatment of the disease, there are a number of caveats that limit the effectiveness of these treatments alone. First of all, surgery is obviously not an applicable therapy for blood cancers. In addition, surgical dissection and/or radiation therapy for the treatment of solid tumors cannot always ensure the removal of every cancer cell, even with the assistance of advanced imaging technologies. Moreover, if the cancer has metastasized (see **metastasis**) or spread to multiple areas of the body away from the primary tumor[1], surgery and/or

---

1 Cancers can be classified based upon their location in the body relative

radiation therapy alone may be of limited benefit. For these reasons, additional treatments involving the use of chemicals (**chemotherapy**) are now commonplace in the treatment of cancer.

Standard chemotherapeutic drugs work by killing cancer cells

The idea of using chemicals to treat cancer was initially met by significant resistance from early cancer therapists (predominantly surgeons) and for good reason. The chemicals initially proposed for the treatment of cancer were also highly toxic to normal cells, *i.e.*, they were considered by most early cancer therapists as poisons and not suitable for human consumption. For example, mustard gas was a chemical initially used to attack soldiers fighting in the trenches of the World War I (Figure 8). The fact that soldiers (and innocent civilians) who managed to survive these lethal attacks were often found to display a significant depletion of their white blood cells (**lymphocytes**), suggested to early advocates of chemotherapy that nitrogen mustard (a chemical derivative of the sulfur mustard gas used in World War I)

---

to their tissue of origin. When a tumor is small, confined to the tissue of origin and has not yet spread or metastasized, it is classified as Stage I. Somewhat larger tumors that remain confined to the tissue of origin (or immediately adjacent) are classified as Stage II. Stage III tumors are those that have displayed limited spread to other tissues (including lymph nodes) while Stage IV tumors are those that have spread extensively throughout the body.

might be an effective treatment for **lymphoma**. Lymphoma is a cancer characterized by overproduction of white blood cells. It turns out that this was a correct assumption and, in fact, the initial success of this chemical treatment against lymphomas ultimately led to nitrogen mustard becoming the first cancer drug approved by the **Federal Drug Administration (FDA)** in 1949.

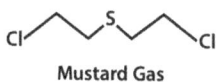

Figure 8- **Mustard Gas**. A chemical initially used to attack soldiers fighting in the trenches of the First World War (Everett Collection/Shutterstock.com).

During the 1950s, a number of other chemicals were identified as being effective against the growth of one or more types of tumors, and many of these were subsequently also approved of as cancer drugs by the FDA. In the 1960s, the efficacy of treating patients with various combinations of drugs was intensively pursued, and such combination treatments were sometimes found to result in dramatic responses. For example, **combination drug therapies** developed in the 1960s were successful in raising the survival rate for patients with the blood cancer, **Hodgkin's lymphoma,** from zero to over 70%.

Although the term, "chemotherapy," was initially coined in the early 1900s in reference to the general use of chemicals to treat any disease, today it is almost exclusively used in reference to cancer drugs (chemicals) that act by interfering with the ability of cells to divide or replicate. Since the most distinguishing characteristic of nearly all cancer cells is rapid and uncontrolled cell division, these drugs (**cytotoxic drugs**) are intended to attack and kill rapidly dividing cancer cells. However, other rapidly dividing cells in our body (*e.g.*, blood cells, hair cells, cells lining the intestine, *etc.*) are also potential targets. For this reason, significant negative side effects can be associated with the administration of this class of cancer drugs (Figure 9). Indeed, one of the major challenges in the administration of chemotherapeutic drugs is finding a dosage level that effectively kills cancer cells while minimizing damage to non-cancer cells.

Chemotherapeutic drugs remain in widespread use today. They are classified based on their chemical structure and specific mechanisms by which they block the cellular machinery underlying cell replication (Table 1). Although the details may vary, all chemotherapeutic drugs cause irreparable cellular (especially DNA) damage to rapidly dividing cancer cells. As is the case with

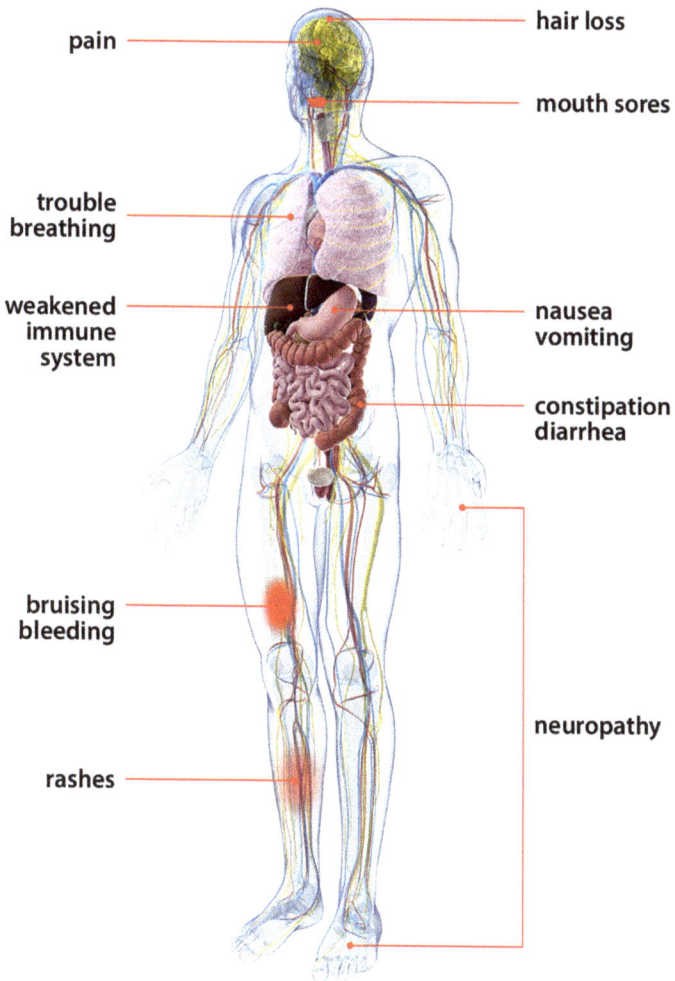

Figure 9- **Chemotherapy (cytotoxic) drugs** have been associated with a variety of negative side effects. The intensity of side effects may vary between individuals and also may depend upon the particular drug and dosage level prescribed. Shown are the 10 most common side effects of cytotoxic chemotherapy (SciePro/Shutterstock.com).

radiation therapy, such extensive cellular/DNA damage typically results in the induction of apoptosis or programmed cell death. This is the primary mode of action of this class of cancer drugs. Because chemotherapeutic drugs are typically delivered intravenously (although some can be administered orally), they become widely distributed throughout the body (*i.e.*, **systemic drug delivery**), which explains why non-cancerous rapidly dividing cells are also prone to attack.

Table 1. Examples of chemotherapeutic drugs classified based on their mechanism of action.

Alkylating Agents	Antimetabolites	Anti-tumor antibiotics
Busulfan	Azacitidine	Doxorubicin
Cisplatin	Cytarabine	Idarubicin
Mechlorethamine	Fludarabine	Bleomycin
Oxaliplatin	Methotrexate	Mitomycin-C
Trabectedin	Pralatrexate	Mitoxantrone

Corticosteroids	Mitotic Inhibitors	Nitrosoureas
Dexamethasone	Docetaxel	Carmustine
Methylprednisolone	Paclitaxel	Lomustine
Prednisone	Vinblastine	Streptozocin
	Vincristine	
	Vinorelbine	

Other		Topoisomerase Inhibitors
Asparaginase		Irinotecan
Hydroxyurea		Topotecan
Mitotane		Etoposide
Pegaspargase		Mitoxantrone
Vorinostat		Teniposide

Hormone therapy may slow or stop the growth of cancers that use hormones to grow

A second class of cancer drugs acts by blocking the induction of cancer cell division/replication by **hormones** (Table 2). For example, even moderate levels of **estrogen** in women with breast cancer and **androgen** in men with prostate cancer has been found to be associated with enhanced tumor growth. **Hormone therapy** involves the administration of drugs that either block the ability of cancer patients to produce specific hormones or interfere with the action of specific hormones on cancer cells. Since hormone therapy drugs are typically administered either orally or by injection, they are also widely (systemically) distributed throughout the body. For this reason, hormonal therapy, like chemotherapy, can also be associated with negative side effects.

Table 2. Hormone therapy is used to treat cancers that need hormones to grow. Below are some examples of hormone therapies used to treat prostate and breast cancers.

Cancer	Type of Hormone Therapy	Examples of Drugs	How they Work
Breast	Estrogen antagonists	Fulfestrant Raloxifene Tamoxifin Toremifene	Block estrogen receptors from interacting with estrogen
	Estrogen synthesis inhibitors	Anastrozole Exemestane Letrozole	Block estrogen production by inhibiting the enzyme aromatase
Prostate	Anti-Androgen	Casodex Eulexin Nilandron	Block testosterone interaction with cancer cells
	Luteinizing hormone-releasing hormone (LHRH) agonists	Eligard Lupron Trelstar Viadur Zoladex	Inhibit the release of LHRH involved in testosterone production in testes
	Blockers of androgen receptors	Apalutamide Erleada Nubeqa Xtandi	Block the interaction of growth stimulating hormones with cells

Targeted therapies inhibit specific proteins contributing to cancer onset and progression

A third class of cancer drugs, collectively referred to as **targeted therapies**, has emerged in recent years thanks to a growing understanding of the specific molecular processes underlying the disease. In contrast to the generalized cytotoxic effects associated with traditional chemotherapeutic drugs, targeted therapies are small inhibitory molecules or antibodies designed to target and inhibit the activity of *specific* proteins, or cellular processes, contributing to cancer onset and progression. Since, as discussed in Chapter 1, there are alternative molecular pathways leading to cancer, the specific protein(s) targeted for therapy by this class of drugs may be expected to vary, not only among patients diagnosed with different types of cancer but even between individual cancer patients with the same clinical diagnosis.

The potential to personalize cancer treatment by targeted therapies (also called, **personalized or precision cancer therapy**) is wholly dependent upon an ability to identify the specific gene or genes disrupted in individual patient tumors. This ability only became a possibility with the advent of technologies enabling the **DNA sequencing** and/or **gene expression profiling** of individual patient tumors. For example, the cancer cells of essentially all **chronic myelogenous leukemia (CML)** and many **acute lymphoblastic leukemia (ALL)** patients were long known to display an abnormal chromosome configuration (called the **Philadelphia Chromosome**). Human DNA is packaged in 23 pairs of chromosomes, each of which is easily distinguishable under a microscope. When a laboratory-produced image of a person's chromosomes isolated from an individual cell is displayed in numerical order (*i.e.,* 1-23) it is called a **karyotype**. The Philadelphia Chromosome involved a reciprocal transfer of segments (translocations) between chromosomes, 9 and 22. (Figure 10). The sequencing of DNA isolated from the cancer cells of CML patients revealed that this chromosomal translocation resulted in a gene called **BCR** being overexpressed because the translocation event repositioned it next to a highly expressed gene called ABL. The resultant mutant **fusion gene** is called **BCR-ABL** and its overexpression results in uncontrolled cell proliferation. Once this target gene was identified, a small molecule called, **imatinib** (brand name Gleevec) that could bind to the BCR-ABL mutant protein and inhibit its activity was developed. Imatinib became a highly effective targeted therapeutic for

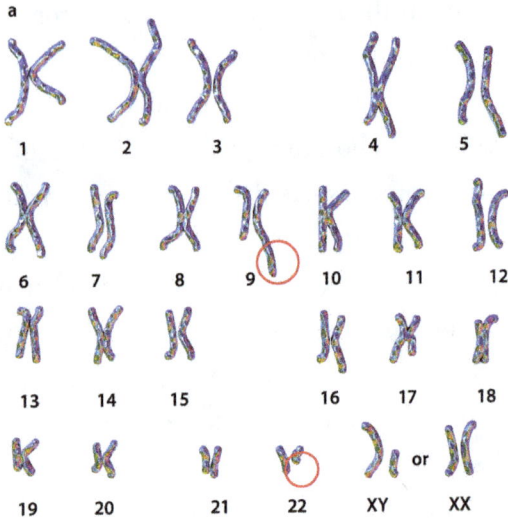

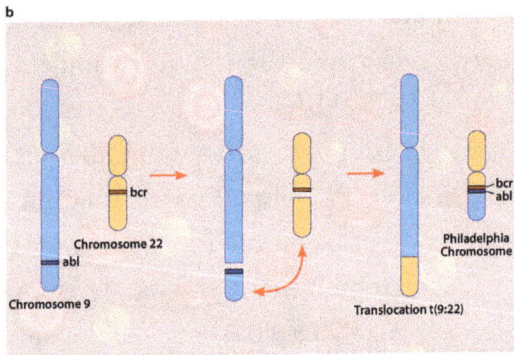

Figure 10- The Philadelphia Chromosome. a) Karyotype of the Philadelphia Chromosome showing the reciprocal transfer of segments (translocations) between their chromosomes 9 and 22. This chromosomal abnormality is associated with the cancer cells of essentially all chronic myelogenous leukemia (CML) and many acute lymphoblastic leukemia (ALL) patients. b) The sequencing of DNA isolated from the cancer cells of CML patients revealed that this chromosomal translocation resulted in a gene called BCR (BCR is normally involved in the highly regulated control of cell replication) being fused to another highly expressed gene (ABL). Over-expression of the mutant fusion gene (BCR-ABL) results in uncontrolled cell proliferation (Kateryna Kon/Shutterstock.com; Meletios Verras/Shutterstock.com).

CML, as well as, for other leukemias displaying a similar disruption in BCR expression (Figure 11).

The BCR-ABL Targeted Drug Imatinib (Gleevec) Blocks Cancer Cell Proliferation

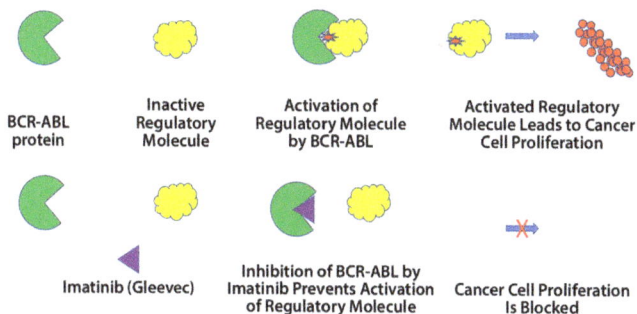

Figure 11- **The BCR-ABL Targeted Drug Imatinib (Gleevec) Blocks Cancer Cell Proliferation.** The BCR-ABL protein activates a regulatory molecule that stimulates cell division (proliferation). Imatinib binds to the BCR-ABL protein preventing it from activating the regulatory molecule thus blocking cancer cell growth.

Today, there are a number of inhibitors available that are capable of targeted inhibition of a variety of abnormal, or abnormally expressed, proteins believed to be involved in cancer onset/development (Table 3). The selection of the appropriate drug or drug combination by clinicians is guided by genomic profiling of individual patient tumors.

Table 3. Many small-molecule targeted cancer therapy drugs have been developed over the last two decades (only a few examples are listed; see https://www.cancer.gov/about-cancer/treatment/types/targeted-therapies/approved-drug-list for a more extensive list by cancer type).

Small Molecule Drug	Cancer Targeted
Imatinib (Gleevec)	Chronic myelogenous leukemia
Erlotinib (Tarceva)	Non-small cell lung cancer, pancreatic cancer
Bevacizumab (Avastin)	Colon, lung kidney, brain, cervical, ovarian cancers
Cetuximab (Erbitux)	Colorectal, head and neck cancers
Sorafenib (Nexavar)	Late stage kidney cancer
Lapatinib (Tykerb)	EGFR, HER2/neu Breast cancer
Vandetenib (Caprelsa)	Medullary thyroid cancer
Ipilimumab (Yervoy)	Melanoma, kidney, liver, lung, colorectal, esophageal

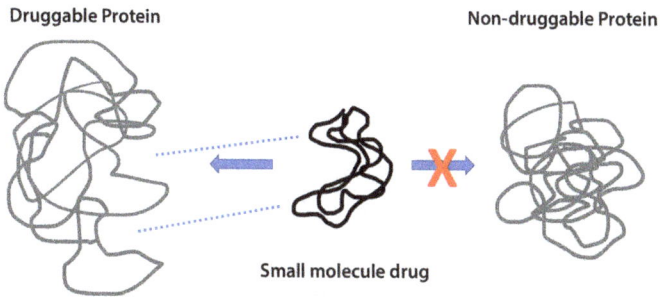

Figure 12- Druggable vs Not Druggable proteins. Not all proteins have a configuration (shape/structure) that makes them conducive to inhibition by small molecules binding to regions critical to protein function.

While targeted therapies are certainly a significant addition to the arsenal of drugs currently available for the treatment of cancer, they are not without their limitations. First, not all proteins have a configuration (shape/structure) that makes them conducive to inhibition by small molecules (Figure 12). Indeed, it is currently estimated that only about 10-15% of human proteins are "druggable" by small inhibitory molecules. This suggests that not all (and perhaps only a minority) of proteins identified as contributing to cancer will be able to be inhibited by targeted therapies. A second limitation is the fact that the structures of targeted cancer-driver proteins that make them accessible and conducive to inhibition by a specific drug are often displayed by other proteins as well. Therefore, a drug that is designed to inhibit a specific cancer driver protein may fortuitously inhibit one or more other proteins as well, resulting in undesirable negative side effects. As a general rule, however, the negative side effects associated with targeted therapies are less than those associated with the more broadly toxic chemotherapeutic drugs.

Immunotherapy enhances the ability of our immune system to recognize and destroy cancer cells

Our **immune system** is designed to identify and destroy foreign substances that may enter our body. Often these substances are foreign proteins or other molecules carried on the surface of pathogenic bacteria or viruses but, in principle, any substance that is not produced by our own body will be identified and attacked by our immune system. During our embryonic development, and soon after birth, our immune system is programmed to recognize (and ignore) proteins and other molecules that we naturally produce in order to prevent us from destroying our own cells. The molecular processes underlying the ability of our immune system to distinguish "self from non-self" are varied and complex but often involve the production of molecules by our normal cells that can effectively block the immune response.

Cancer cells often produce and display mutant proteins and other aberrant molecules on their surface that appear to be "foreign" to our immune system. Indeed, it is believed that many cancerous and pre-cancerous cells arising in our body are destroyed by our immune system on a daily basis before we even know they exist (Figure 13).

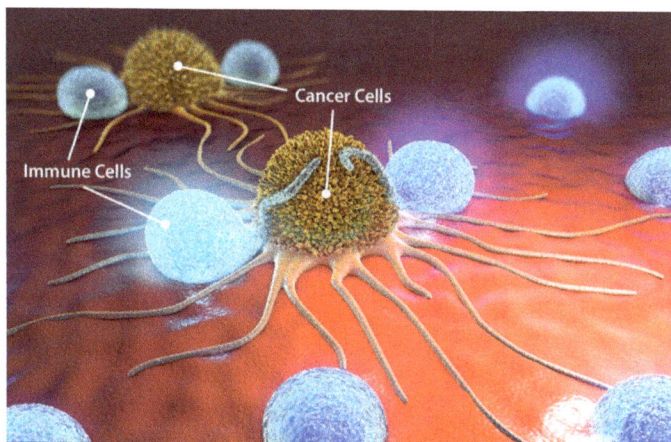

Figure 13- Destruction of Cancer Cells by Immune Cells. Many cancerous and pre-cancerous cells are believed to be destroyed by our immune system early in cancer development. Cancer cells often produce and display mutant proteins and other aberrant molecules on their surface that appear to be "foreign" and thus attacked by our immune system (Christoph Burgstedt/Shutterstock.com).

Not all cancer cells, however, display sufficiently aberrant molecules on their surface such that they are recognized as foreign by our immune system. In addition, as we grow older our immune system (and unfortunately many of our other body parts and functions as well!) begin to deteriorate. This breakdown increases the probability that cancer cells may escape immune surveillance and begin to grow unchecked in our bodies. Indeed, this is believed to be one of the major reasons why the frequency of cancer is significantly higher in older people rather than in younger people (Figure 14). Another mechanism by which cancer cells (even those displaying foreign molecules on their surface) may escape death is by aberrantly overproducing molecules used by our normal cells to block the immune response.

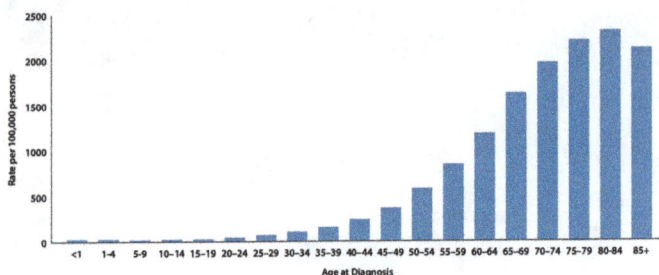

Figure 14- **Cancer Frequency Increases with Age.** Advancing age is the most important risk factor for many individual cancer types. The overall incidence rates for cancer tend to increase as we age.

Immunotherapy encompasses a variety of methods and technologies all designed to enhance the ability of our immune system to recognize and destroy cancer cells (Table 4). For example, one strategy is to inhibit those molecules overproduced by cancer cells that block the immune response (Figure 15). This is accomplished by treating patients with drugs (specific protein inhibitors or highly specific antibodies) designed to bind to and inhibit these immune response blocking molecules (**immune checkpoint inhibitors**). However, since these immune system blocking molecules are also produced by our normal cells, patients treated with these "checkpoint inhibitors" need to be carefully monitored. Treatment levels have to be adjusted, if necessary, to avoid drug-induced **auto-immunity** (destruction of the body's normal cells by our own immune systems).

Table 4. Five major types of cancer immunotherapy.

Cancer Immunotherapy	How It Works	To Treat
Cell Therapy	Modifies immune cells to fight cancer	CAR-T for leukemias, lymphomas
Checkpoint Inhibitors	Prevents tumor from turning off cancer-fighting cells	Melanoma, Hodgkin lymphoma, Merkel cell, cutaneous squamous cell carcinoma, head and neck cancer, triple negative breast cancer, lung, colorectal, kidney, bladder, and others
Cytokines	Boosts the immune system	Melanomas, kidney cancers
Oncolytic Virus Therapy	Uses viruses to fight cancer	Advanced melanomas
Vaccines	Teaches the body's immune cells to find cancer cells	Prostate cancer

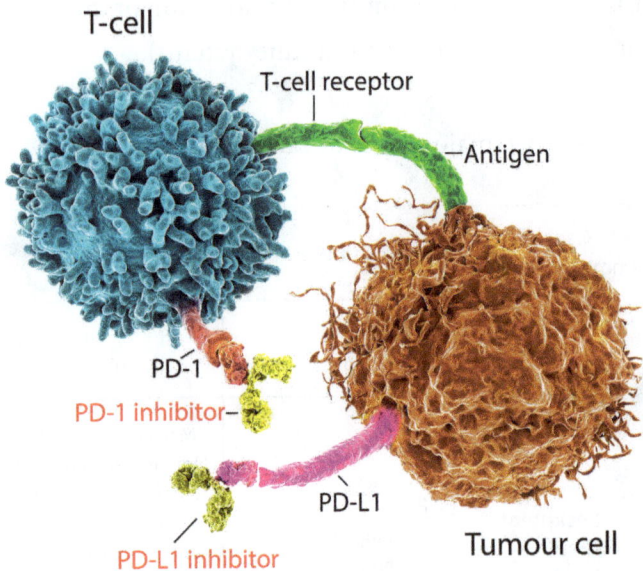

Figure 15- Immunotherapy by inhibition of Check Point Inhibitors. Our normal cells produce molecules, called checkpoint inhibitors (*e.g.*, PD-L1), on their surface that prevents our immune cells (T-cells) from destroying our own cells. Cancer cells often over-express these checkpoint inhibitors (*e.g.*, PD-L1) to prevent their recognition (*e.g.*, by PD-1) by T-cells and their consequent destruction by our immune system. Drugs (*e.g.*, inhibitors against PD-L1 or PD-1) block this immune suppressive strategy making cancer cells more susceptible to destruction by immune cells (Kateryna Kon/Shutterstock.com).

A variety of other immunotherapies are currently in use or under development for the treatment of various cancers. In some cases, these treatments may simply involve the administration of drugs (called **cytokines**) designed to generally enhance the immune response. More intensive therapies may include the removal of immune cells from cancer patients in order to grow and expand their numbers in the laboratory before reintroduction into the patient. In some cases, the collected immune cells may not only be expanded in numbers in the laboratory but also genetically modified/engineered in ways to enhance their cancer killing ability before being reintroduced into the patient (*e.g.*, **CAR-T Therapy**) (Figure 16).

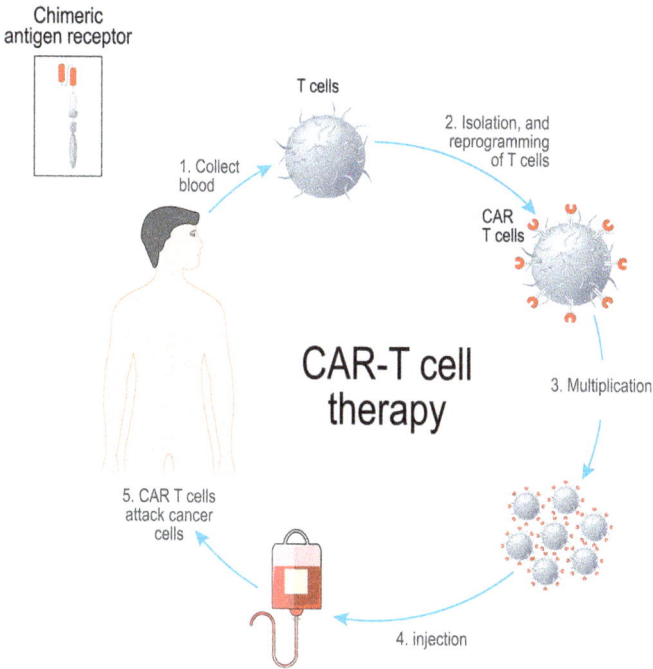

Figure 16- CAR-T Cell Therapy. In this procedure, immune cells (T-cells) are removed from the blood of a cancer patient and "engineered" (*i.e.*, the addition of molecules called Cancer Antigen Receptors or CARs) to enhance the ability of the T-cells to recognize and attack the patient's specific cancer when reintroduced into the bloodstream (Designua/Shutterstock. com).

The general idea of stimulating our body's defense system(s) to recognize and kill cancers is intuitively appealing and has great potential. However, at present, immunotherapy is still a highly evolving field, and there is still much to be learned. For some cancer patients the positive impact of current immunotherapies has been dramatic while for others they have been of limited benefit. Indeed, a major focus of current immunotherapy research is focused on the development of methods to identify those cancer patients most likely to respond to currently available treatments. Despite these challenges, there is no question that immunotherapy holds great future promise in the treatment of cancer.

Viruses can be designed to infect and kill cancer cells

As early as the late 1800s, there were sporadic reports of cancer patients going into at least temporary cancer remission following viral infections. Subsequent early efforts to select for infectious viruses having a propensity for killing cancers were generally unsuccessful because the infecting viruses were usually recognized and destroyed by the patients' immune system before they could successfully attack cancer cells. Modern improvements in methods to genetically modify cancer killing viruses has led to a renewed interest in the potential use of viruses in cancer therapy.

In recent years cancer killing viruses (today, called **oncolytic viruses**) have been genetically engineered to specifically bind to and replicate within cancer cells. When these viruses replicate, the resulting progeny viruses burst open the host cancer cells destroying them in the process. As progeny viruses are released, the intent is that they will go on to infect adjacent cancer cells and/or release viral antigens recognized as foreign by our immune system resulting in a chain reaction of cancer cell deaths. While more research will be required before the methodology is perfected, scientists are hopeful that oncolytic viruses will, in the not-too-distant future, become a welcome addition to the arsenal of newly developed approaches to cancer therapy.

Summary

There are a variety of effective tools and procedures currently available for the treatment of cancer and many promising new treatments are currently under development. As a general rule, none of these alternative strategies are mutually exclusive and, indeed, most cancer treatments are often most effective when used in combination.

Additional References and Links

Robotic Surgery
What is the Difference Between Traditional and Robotic Surgery?, Columbia University video:
https://www.youtube.com/watch?v=Q7lmB2FIfdo

Da Vinci Robotic Surgery Program. Cleveland Clinic video:
https://www.youtube.com/watch?v=YVWgRjWfD9k

Radiation Therapy

What is Cancer Radiotherapy and How Does It Work? Cancer Research UK video:
https://www.youtube.com/watch?v=V2VGHUjN17w

How Radiotherapy Works. GenesisCare UK video:
https://www.youtube.com/watch?v=3_pdhKoWu7I

Brachytherapy. University of Rochester video:
https://www.youtube.com/watch?v=csamOxq4DaA

Chemotherapy

How to Treat Cancer. 7activestudio video:
https://www.youtube.com/watch?v=GyiobIq4fnk

Hormonal Therapy

Beyond The Shock - Treatment - Hormone Therapy, National Breast Cancer Foundation video:
https://www.youtube.com/watch?v=F4cVLuENaOk

What is Hormone Therapy (prostate cancer), Prostate Cancer UK video: https://www.youtube.com/watch?v=F1iSnaT6nF8

Targeted Therapies

Targeted Therapy for Cancer. Samitive Hospitals video:
https://www.youtube.com/watch?v=KL3sktEq4KM

What are Targeted Therapies for Cancer Treatment? Cancer. Net video:
https://www.cancer.net/navigating-cancer-care/videos/treatments-tests-and-procedures/what-are-targeted-therapies-cancer-treatment

Immunotherapy

The Path to a Cancer Cure video:
https://www.youtube.com/watch?v=UbFjiWOBErA

Car-T Therapy: How Does It Work? Dana Farber Cancer Institute video:
https://www.youtube.com/watch?v=OadAW99s4Ik

Oncolytic Viruses

Oncolytic Virus Therapy: Dynamite for Cancer Cells. Cancer Research Institute video:
https://www.youtube.com/watch?v=zwlCkVnUgWQ

Can Viruses Kill Cancer. CBS News: The National video:
https://www.google.com/search?q=cancer+killing+viruses&rlz=1C5GCEM_enUS1029US1031&source=lnms&tbm=vid&sa=X&ved=2ahUKEwjF-srmyu37AhWiY98KHQZFAqYQ_AUoA3oECAMQBQ&biw=1280&bih=639&dpr=1#fpstate=ive&vld=cid:1e600631,vid:-AHsFKK9YLE

Chapter 4
Early Cancer Diagnostics

Early cancer diagnosis is important for recurrence as well as for the onset of the disease

While this book was written primarily for individuals already diagnosed with cancer, early detection of recurrence of the disease in previously treated patients makes the topic of cancer diagnostics highly relevant not only to the general population but to cancer patients as well. As a general rule, any effective diagnostic test needs to be cost-effective (an especially relevant issue for insurance companies) and minimally invasive (an especially relevant issue from the perspective of the person undergoing the test). However, the most important criterion overall is accuracy.

Highly accurate cancer diagnostic tests have both high sensitivity and high specificity

Ideally, a cancer diagnostic test should not only be able to accurately detect the disease when it is present (high **sensitivity**) but also be able to accurately rule out the disease when it is not present (high **specificity**). Currently, the most accurate diagnostic tests are, relatively speaking, the most invasive and costly to perform. Hence, they are usually reserved for those individuals considered to be at an elevated risk. For example, previously treated cancer patients currently in partial or complete remission are, nevertheless, at increased risk of (re-) developing cancer. For this reason, they are typically screened annually or bi-annually, oftentimes by one of the highly accurate radiological procedures discussed in Chapter 3 (*e.g.*, X-rays, CT, PET scans or, MRIs). The goal is to detect any evidence of cancer recurrence early on when treatment is most effective.

The need for diagnostic screening increases as we age

As discussed briefly in Chapter 3, advancing age alone is associated with an, at least, moderate increase in the risk of developing cancer. Because the elevated risk of developing cancer as we grow older is typically not as high as the risk of recurrence for previously treated cancer patients, the more intensive and typically more costly procedures are not routinely recommended nor employed for older people unless suspicious pre-clinical symptoms are present. One exception is **mammograms** (X-rays of the breast) that are recommended for all women above the age of 50 at least once every two years (Figure 17). Although moderately costly, routine mammograms are typically covered by most insurance programs because early diagnosis of breast cancer is not only medically beneficial but cost-effective as well.

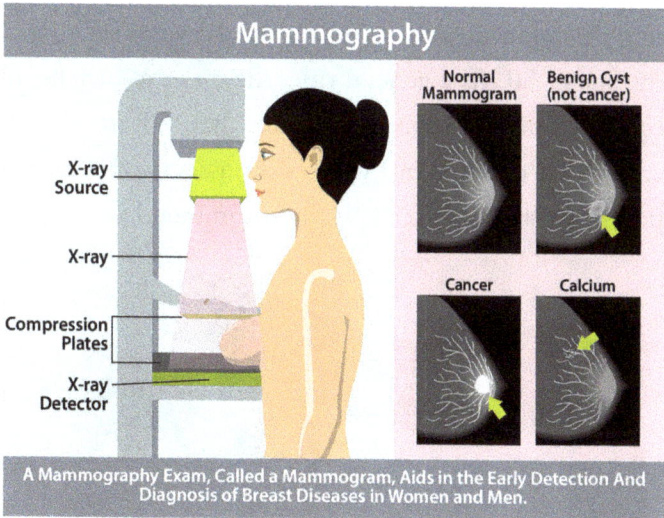

A Mammography Exam, Called a Mammogram, Aids in the Early Detection And Diagnosis of Breast Diseases in Women and Men.

Figure 17- **Mammography.** A specific type of breast imaging that uses low-dose X-rays passing through breast tissue and interacting with a digital detector to see inside the breasts. A mammography exam, called a mammogram, aids in the early detection and diagnosis of cancer and other breast diseases in women (Luis Line/Shutterstock.com).

For seemingly healthy older individuals, other types of effective diagnostic screenings are also often recommended. For example, **colonoscopy** is a procedure that utilizes an endoscope to allow a physician to examine the rectum and colon for polyps (Figure 18). If detected, the polyp(s) is (are) removed and examined in the laboratory for evidence of cancer. Although colonoscopies can hardly be described as non-invasive, they are both medically beneficial and cost-effective because they are currently the best way of detecting colon cancer at an early stage when it can be most successfully treated. Asymptomatic individuals are recommended to have colonoscopies at least every 10 years starting at the age of 50. As a general rule, the medical benefits of colonoscopies in individuals older than 75 are considered to be overridden by the risks associated with the procedure and are not typically recommended.

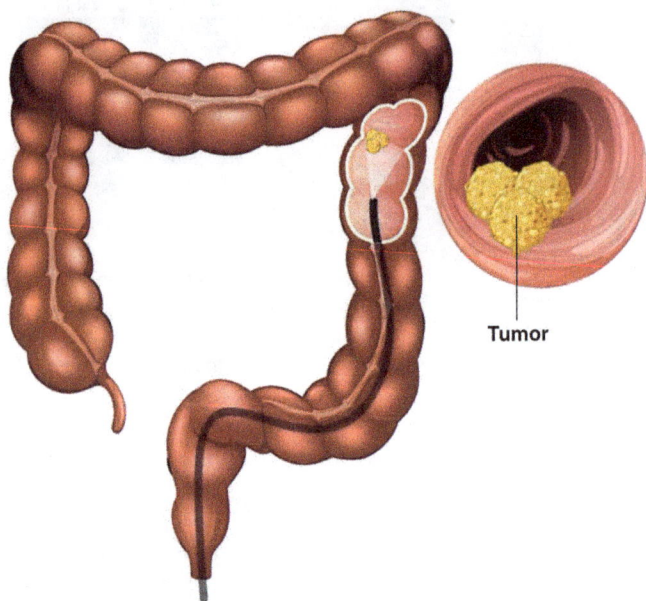

Tumor

Figure 18- Colonoscopy. An exam to look for polyps or early-stage cancer in the large intestine (colon) and rectum. The procedure involves the insertion of a long, flexible tube (colonoscope) into the rectum. The scope inflates your large intestine with air for a better view and contains a camera that sends a video image to a monitor. If necessary, polyps or other abnormal tissue can be removed and biopsies can also be taken (BlueRingMedia/ Shutterstock.com).

Alternative molecular pathways to even the same type of cancer can reduce the accuracy of diagnostic tests

As described above, the ideal diagnostic test should be minimally invasive (using blood or other easily accessible body fluids, such as urine or sputum) and able to detect cancer with high accuracy. Development of such an ideal cancer diagnostic has, thus far, proven to be elusive. The reason for this is, at least in part, the same reason why the development of broadly effective targeted drug therapies has also proven to be difficult (see Chapter 1), *i.e.*, the underlying molecular processes leading to even the same type of cancer can be highly variable. This heterogeneity makes the identification of reliable molecular biomarkers of even the same cancer type difficult to identify, often resulting in tests associated with low specificity.

Molecular indicators of cancer may also be shared by non-cancerous conditions

Another hurdle in the quest for a highly accurate diagnostic test for even the same cancer type is the fact that a number of the secondary molecular processes disrupted in cancer may also be associated with non-cancerous conditions. Any biomarker(s) associated with such processes would not necessarily be indicative of cancer, potentially resulting in "false positives" and reduced specificity. A prime example of this is the **PSA (prostate specific antigen) test** for prostate cancer. While elevated levels of PSA (typically defined as serum levels of 4.0 ng/ml) are often taken as *prima facia* evidence of at least an increased probability of prostate cancer, elevated levels of PSA can also be caused by benign prostate diseases (*e.g.*, benign prostatic hyperplasia, acute/chronic prostatitis, elevated urinary tract infection) or by systemic inflammation. Indeed, short term/transient elevations in PSA levels can be the result of recent sexual activity or other physical activities (*e.g.*, bicycle riding). Such ambiguities become an issue when decisions have to be made by patients/clinicians about possible follow-up procedures, such as invasive biopsies. Indeed, only about 25% of people who have a biopsy done because of an elevated PSA level are ultimately found to have prostate cancer.

Promising new approaches to early cancer diagnosis are currently under development

Despite the challenges outlined above, today, the development of highly accurate, non-invasive cancer diagnostic tests is an area involving intense research activity, and many promising results are beginning to emerge. For example, recently developed tests (sometimes called "**liquid biopsies**") focusing on the detection of cancer cells, DNA or RNA from cancer cells circulating in the blood that are indicative of various types of cancer (*e.g.*, characteristic mutations in cancer driver genes), are showing promising results (Figure 19). Likewise, the recent application of computer-assisted **machine learning** methods to analyze the thousands of proteins and metabolites circulating in our blood are identifying highly significant differences that may, in the not-too-distant future, be used to accurately diagnose both early-stage cancer onset and recurrence.

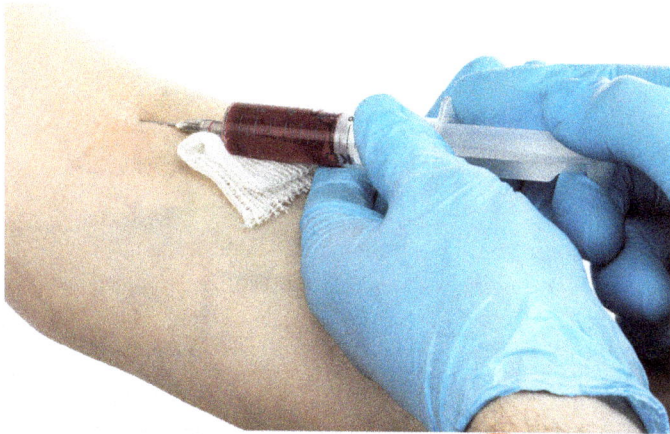

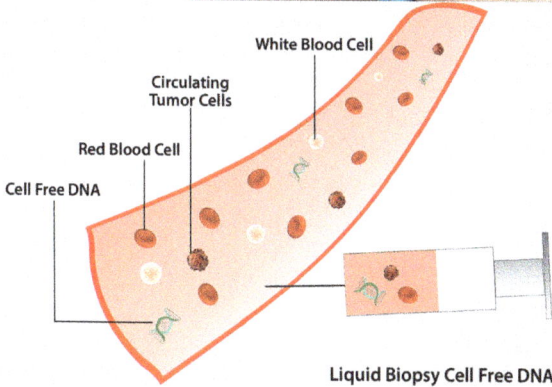

Figure 19- Liquid Biopsy. As a tumor grows, pieces can break off and circulate in your bloodstream. A liquid biopsy is a laboratory test done on a sample of blood, urine, or other body fluid to look for cancer cells from a tumor or small pieces of DNA, RNA, or other molecules released by tumor cells into a person's body fluids (Szaboles Borbely/Shutterstock. com; Abhinav Chaudhary/Shutterstock.com).

Additional References and Links

Readings

How Cancer is Diagnosed. National Cancer Institute (article):
https://www.cancer.gov/about-cancer/diagnosis-staging/diagnosis

How is Breast Cancer Diagnosed?
https://www.cdc.gov/cancer/breast/basic_info/diagnosis.htm

Breast Cancer Screening. National Cancer Institute (article):
https://www.cancer.gov/types/breast/patient/breast-screening-pdq

Prostate Cancer Screening. National Cancer Institute (article):
(https://www.cancer.gov/types/prostate/patient/prostate-screening-pdq

Colon Cancer Screening. National Cancer Institute (article):
https://www.cancer.gov/types/colorectal/patient/colorectal-screening-pdq

Demystifying Liquid Biopsies. American Society of Hematology (article):
https://ashpublications.org/ashclinicalnews/news/3838/Demystifying-Liquid-Biopsies?searchresult=1

Can Artificial Intelligence Help See Cancers in New and Better Ways? National Cancer Institute (article):
https://www.cancer.gov/news-events/cancer-currents-blog/2022/artificial-intelligence-cancer-imaging#:~:text=NCI%20researchers%20have%20built%20and,cervix%20with%20a%20small%20camera.

Videos

How is Prostate Cancer Diagnosed? GoodRX Health video:
https://www.goodrx.com/conditions/prostate-cancer/how-is-prostate-cancer-diagnosed

What Does It Mean to Have Dense Breasts? Centers for Disease Control video):
https://www.cdc.gov/cancer/breast/index.htm

What Happens During and After a Colonoscopy? You and Colonoscopy video:
https://www.youtube.com/watch?v=mh90RPA-C10&t=2s

Chapter 5
Some Final Thoughts

Once the initial shock of being diagnosed with cancer has subsided, most individuals focus on how they might be optimally treated. As you now know, this is not a question with a simple answer. There are, in fact, many excellent treatment options currently available, and the optimal path for each individual patient will depend upon the particular cancer type and the degree to which the cancer has progressed.

As a general rule there are well-established standards of care for most types and stages of cancer, and the majority of cancer clinicians typically follows these consensus treatments. This is not to say that all cancer specialists will necessarily agree as to what they consider to be the optimal treatment for an individual patient. In many cases, the treatment(s) preferred by individual physicians may be determined, in large part, by their particular areas of expertise and experience. This is not a bad thing because when effective alternative treatments are possible, clinical experience is often the most important consideration.

Nevertheless, it is quite reasonable for you, as a patient, to obtain second (or third) opinions from different physicians before you set forth on your personalized therapeutic path. If you do choose to seek second opinion(s), don't be afraid to ask questions and have meaningful discussions with your clinician(s). Not only might you learn something new, these discussions may help you to evaluate the competence of the person or persons you may be entrusting with your life!

Because cancer is such a complex disease and because different cancers often present unique clinical challenges, extensive prior experience, as mentioned above, is often the best and most important credential to look for in selecting a physician. For example, if you have breast cancer, being treated by a specialist who treats a hundred or more breast cancer patients per year may be preferable to being treated by a physician who only occasionally treats the disease. Other relevant factors may include where the physician was trained (*e.g.*, was it a reputable institution providing training in advanced therapies, *etc.*) and the facilities (*e.g.*, is the hospital or clinic where the physician practices equipped with the resources necessary to optimally treat the disease?). However, what is by far most important is that you find a physician and/or clinical team with whom you feel both comfortable and confident. This will help give you the confidence, the determination and the fortitude needed to successfully take on the challenge facing you.

As we have seen, highly effective treatments are already available, and future (even short term) prospects are looking extremely good. We can reasonably expect that minimally invasive surgical techniques, coupled with enhanced imaging technologies, will continue to improve over the next several years. Another highly promising area of cancer treatment that is only beginning to emerge is **cancer immunotherapy.** We can confidently expect the field of cancer immunotherapy to rapidly expand in the near future with widespread benefits in the treatment of all cancer types. Likewise, the fields of precision medicine and targeted drug therapies are poised to expand dramatically in the near future as scientists acquire an ever more detailed understanding of the alternative molecular pathways leading to cancer in individual patients. New methodologies are rapidly developing to deliver cancer drugs (sometimes contained within microscopic **nanoparticles**) directly to tumors allowing the employment of more effective therapies while avoiding the negative side effects typically associated with the system-wide delivery of cancer drugs.

Just as we have recently witnessed a rapid and dramatic revolution in how viruses (most notably COVID) can be more effectively treated by mRNA-based, rather than traditional protein-based targeted therapies, so too can we anticipate that RNA-based therapies will be developed to compliment and possibly help improve upon protein-based cancer therapies (*e.g.*, targeted protein antibodies and inhibitors). In addition, as mentioned above, recent advancements in the development of nano-technologies allowing targeted therapies to be precisely, rather than systemically, delivered to cancers promises to significantly reduce, if not eliminate, many of the negative side effects traditionally associated with cancer chemotherapy.

The integration of various aspects of computer science into the field of cancer biology is already having a significant impact and this will most certainly continue and expand in the future. In addition to the continuing improvements in computer-assisted imaging technologies mentioned above, we can anticipate ever-expanding applications of **artificial intelligence (AI)** and **machine learning (ML)** in the field of cancer biology. This will lead to improved personalized diagnostics and the more accurate prediction of optimal personalized drug therapies for individual cancer patients.

Today, a diagnosis of cancer is very far from the mandatory death sentence it was just a few years ago. Rather, there is every reason for hope and optimism that we are looking towards a near future when the vast majority of cancers can and will be effectively controlled.

Glossary

Acute lymphoblastic leukemia (ALL) – a type of cancer in which the bone marrow makes too many lymphocytes (a type of white blood cell).

Androgen – a male sex hormone (see **Hormones**).

Apoptosis – the death of cells that occurs as a normal and controlled part of an organism's growth or development (also called programmed cell death).

Artificial Intelligence (AI) – the theory and development of computer systems able to perform tasks that normally require human intelligence. AI combines computer science and robust datasets to enable problem-solving.

Auto-immunity – the system of immune responses of an organism against its own healthy cells, tissues and other normal body constituents. Any disease resulting from this type of immune response is termed an "autoimmune disease." Examples include celiac disease, Type I diabetes, lupus, and rheumatoid arthritis, among many others.

BCR gene (breakpoint cluster region) – BCR is one of the two genes in the BCR-ABL fusion protein, which is typically associated with some types of blood cancer (see **CML** and the **Philadelphia Chromosome**).

Brachytherapy – a radioactive implant is put inside the body in or near the tumor (also called internal radiation).

Cancer driver genes – genes that significantly contribute to, or "drive" the onset and/or progression of cancer.

Cancer immunotherapy – a type of cancer treatment that uses a person's own immune system to fight cancer (see **immune system**).

CAR-T Therapy – Chimeric antigen receptor (CAR) T-cell therapy is a way to enhance the ability of a patient's immune cells to fight cancer by first isolating, then modifying them in the lab so they can better find and destroy cancer cells. The modified T-cells are then re-introduced into the patient's body.

Cell – the basic building blocks of all living things. The human body is composed of trillions of cells. Cells provide structure for the body, take in nutrients from food, convert those nutrients into energy, and carry out specialized functions, including the replication of DNA and the production of progeny cells (see **Progeny**).

Chemotherapeutic drugs – powerful chemicals used to kill fast-growing (typically cancer) cells in the body. These drugs often severely damage DNA, inducing programmed cell death (see **Apoptosis**).

Chemotherapy – cancer treatments involving the use of chemicals.

Chromosome – a threadlike structure consisting of DNA and protein found in the nucleus of most living cells, carrying genetic information.

Chronic myelogenous leukemia (CML) – an uncommon type of cancer of the bone marrow that causes an increased number of white blood cells in the blood (see **Philadelphia Chromosome**).

Combination drug therapies – simultaneous use of multiple drugs or other agents as part of a treatment regimen against a specific disease to improve effectiveness, prevent the development of drug resistance, and minimize side effects.

Colonoscopy – an exam to look for changes, such as swollen, irritated tissues, polyps or cancer, in the large intestine (colon) and rectum. During a colonoscopy, a long, flexible tube (colonoscope) with a tiny camera is inserted into the rectum.

Computerized tomography (CT) scan – a type of scan that combines a series of X-ray images taken from different angles around the body. It uses computer processing to create cross-sectional images (slices) of the bones, blood vessels and soft tissues inside the body. CT scan images provide more detailed information than plain X-rays do.

Cytokines – any of a number of substances, such as interferon, interleukin, and growth factors, that are secreted by certain cells of the immune system and have an effect on other cells.

Cytotoxic drugs – a drug that damages or destroys cells that is used to treat various types of cancer.

DNA (deoxyribonucleic acid) – the hereditary material in humans and almost all other organisms. Information is stored in DNA as a code made up of four chemical bases: adenine (A), guanine (G), cytosine (C), and thymine (T). Human DNA consists of about 3 billion bases. More than 99 percent of those bases are the same in all people. The order, or sequence, of these bases determines the information available for building and maintaining an organism, similar to the way in which letters of the alphabet appear in a certain order to form words and sentences.

DNA repair genes – genes that encode the proteins involved in the repair of DNA mutations.

DNA repair mechanisms – a means by which cells fix any mutated or otherwise damaged DNA, thus reducing the number of disrupted cells and thereby maintaining normal cell function.

DNA sequencing – determines the precise order of the four chemical building blocks, or "bases," that make up the DNA molecule (see **DNA**). Human cells have about six billion bases. Errors (see **Mutations**) in the bases can cause a cell to malfunction in a variety of ways, including the formation of cancer.

Electron beam therapy – a type of radiation therapy that is not able to penetrate tissues as deeply as photon therapy. It is typically used only to treat cancers on or close to the surface of the skin (see **Photon Beam Therapy** and **Proton Beam Therapy**).

Estrogen – any of a group of steroid hormones that promote the development and maintenance of female characteristics of the body (see **Hormones**).

External beam radiation – the most common type of radiation therapy in current clinical practice. It involves the aiming of high-energy particles generated from a radiation source outside the body to the location of the tumor.

False positive – a test result that indicates a person has a specific disease or condition when the person actually does not have it.

False negative – a test result that indicates a person does not have a disease or condition when the person actually does have it.

Federal Drug Administration (FDA) – the government agency responsible for protecting the public health by assuring the safety, efficacy, and security of human and veterinary drugs, biological products, medical devices, our nation's food supply, cosmetics, and products that emit radiation.

Gene expression profiling – measures RNA levels, showing the pattern of genes expressed by a cell at the transcription level.

Gene – segment of DNA that encodes the information needed to produce a single mRNA molecule used in the synthesis of a single protein.

Genome – all of the DNA we carry in our nuclei (see **Nucleus**).

Hodgkin's lymphoma – a type of cancer in which white blood cells (lymphocytes) grow out of control, causing swollen lymph nodes and growths throughout the body.

Hormones – a product of living cells that circulates in body fluids (such as blood) and produces a specific, often stimulatory, effect on the activity of cells (see for example: **Androgen** and **Estrogen**).

Hormone therapy – a cancer treatment that slows or stops the growth of cancers that use hormones to grow.

Imatinib (brand name Gleevec) – an oral chemotherapy medication used to treat cancer. Specifically, it is used for chronic myelogenous leukemia (CML) and acute lymphoblastic leukemia (ALL) that are Philadelphia Chromosome positive, as well as similar types of cancer.

Immune checkpoint inhibitors – immune checkpoints are a normal part of the immune system. Their role is to prevent an immune response from being so strong that it destroys healthy cells in the body. Immune checkpoint inhibitors work by blocking checkpoint proteins allowing the T-cells to kill cancer cells.

Immune system – a network of biological processes that protects an organism from diseases. It detects and responds to a wide variety of pathogens, such as viruses, parasitic worms, *etc.*, as well as cancer cells, and can distinguish them from the organism's own healthy tissue.

Internal radiation – a radioactive implant is put inside the body in or near the tumor (also called **Brachytherapy**).

Ionizing radiation – radiation with sufficient energy to cause electrically charged particles (ions) to form that can severely damage the DNA of the cells comprising tissue that the radiation passes through.

Karyotype – a laboratory-produced image of a person's chromosomes isolated from an individual cell and displayed in numerical order (*i.e.*, 1-23).

Laparoscope – a fiber optic instrument inserted through the abdominal wall to view the organs in the abdomen (see **Laparoscopy**).

Laparoscopy – a surgical procedure in which a fiber-optic instrument is inserted through the abdominal wall to view the organs in the abdomen or often to assist in a surgical procedure (laparoscopic surgery).

Liquid Biopsy – a laboratory test done on a sample of blood, urine, or other body fluid to look for cancer cells from a tumor or small pieces of DNA, RNA, or other molecules released by tumor cells into a person's body fluids.

Loss-of-function mutation – a change in DNA that results in the decreased production of a protein or a protein with impaired function.

Lymphocytes – a type of immune cell (white blood cell) that is made in the bone marrow and is found in the blood and in lymph tissue. The two main types of lymphocytes are B-lymphocytes and T-lymphocytes.

Lymphoma – a cancer characterized by the overproduction of white blood cells.

Machine learning – the use and development of computer systems that are able to learn and adapt without following explicit instructions by using computer programs and statistical models to analyze and draw inferences from patterns in data (considered a subset of AI).

Magnetic resonance imaging (MRI) scan – a medical imaging technique that uses a magnetic field and computer-generated radio waves to create detailed images of the organs and tissues in the body, including cancers if present.

Mammogram – an X-ray picture of the breast.

Messenger RNA (mRNA) – one form of RNA that functions in encoding the message coming from the DNA to form proteins.

Metastasis – refers to cancer that has spread to a different part of the body from where it started. When this happens, the cancer is said to have "metastasized."

Minimally invasive surgery – surgical techniques that limit the size of incisions needed, thereby reducing wound healing time, associated pain, and risk of infection (see **Laparoscopy**).

Molecular biomarkers – molecules that can be used as an indicator of a particular disease state (*e.g.*, see **Prostate Specific Antigen**).

Mutations – changes in DNA that alter the information encoded in or expressed by a gene.

Nanoparticles – microscopic particles that exist in the natural world or created in a lab and classified on their application, such as in diagnosis or therapy (*e.g.*, use in drug delivery).

Non-encoding RNAs – RNAs that do not get translated into proteins but may play a regulatory role in the cell, contributing to the properly controlled transfer of information out from the DNA database.

Nucleus – membrane-enclosed organelle within a cell that contains DNA (packaged in chromosomes). An array of holes, or pores, in the nuclear membrane allows for the selective passage of certain molecules (*e.g.*, such as proteins and RNA) into and out of the nucleus.

Open surgical techniques – the cutting of skin and tissues so that the surgeon has a full view of the structures or organs involved (as opposed to minimally invasive surgery).

Oncolytic viruses – cancer-killing viruses.

Personalized or precision cancer therapy – medical treatments (typically drugs) customized for individual patients based on the molecular processes underlying each individual's cancer.

Philadelphia Chromosome – chromosomes in some cells may swap sections with each other, (called translocations), sometimes resulting in the promotion of cancer. The Philadelphia Chromosome is a section of chromosome 9 that switches places with a section of chromosome 22, creating an extra-short chromosome 22 and an extra-long chromosome 9. This extra-short chromosome 22 was named after the city where it was first discovered in 1960. More recent molecular studies have demonstrated that this translocation event created a new hybrid gene, called BCR-ABL that leads to the overproduction of white blood cells (see **Chronic Myelogenous Leukemia**).

Photon beam radiation – a type of radiation therapy that uses X-rays or gamma rays that come from a special machine. The radiation dose is delivered at the surface of the body and is able to penetrate tissues and focus high levels of radiation on the tumor (see **Electron Beam Therapy** and **Proton Beam Therapy**).

Positron emission tomography (PET) scan – an imaging test that can help reveal the metabolic function of tissues and organs. The PET scan uses a radioactive drug (tracer) to detect the abnormal metabolism characteristic of many cancers.

Primary tumor – uncontrolled growth of abnormal cells typically culminating in a localized solid mass in a tissue of origin.

Progeny cells – a cell copies its genetic material and then divides to make two identical cells, called progeny, offspring or daughter cells.

Programmed cell death – the death of cells that occurs as a normal and controlled part of an organism's growth or development (also called **Apoptosis**).

Prostate specific antigen (PSA) test – a blood test that measures the level of a protein (PSA) often elevated in cancers of the prostate gland.

Proton beam therapy – a type of radiation therapy that uses energy from positively charged particles (protons). Proton therapy may cause fewer side effects than traditional radiation, since doctors can better control where the proton beams deliver their energy (see **Electron Beam Therapy** and **Photon Beam Therapy**).

Replication – whenever a cell divides, the two new progeny or daughter cells contain the same genetic information, or DNA, as the parent cell. The DNA in the parent cell is copied (replicated) so that each new cell has the same DNA as the parent.

Ribosomes – organelles found within all cells that function to make proteins.

RNA (ribonucleic acid) – nucleic acid present in all living cells that has structural similarities to DNA. RNA is essential in various cellular functions, including the regulation of the expression of genes and coding the information needed for protein synthesis.

Robot-assisted laparoscopic surgery (RALS) – when a doctor guides small robotic arms through tiny incisions, usually in the abdomen or pelvis, to do surgery.

Sensitivity – the proportion of diseased people who are correctly identified as "positive" by the test.

Specificity – the proportion of non-diseased people who are correctly identified as "negative" by the test.

Systemic drug delivery – a non-specific method of delivering drugs that can reach all areas of the body.

Targeted therapies – drugs designed to target specific proteins (genes) believed to be driving specific cancers.

Transcription – the process by which a cell makes a RNA copy from a template DNA (typically a gene).

Translation – the process by which a cell makes proteins using the genetic information carried in messenger RNA (mRNA).

Tumor Margins – the edge or border of tissues removed in cancer surgery. The margin is considered "negative" or "clean" when the pathologist is unable to detect cancer cells at the margin of the tissues surgically removed.

www.ingramcontent.com/pod-product-compliance
Lightning Source LLC
Chambersburg PA
CBHW071503210326
41597CB00018B/2675